Curious EnCOUNTers

1 to 13 Forest Friends

Ben Clanton   Illustrated by Jessixa Bagley

little bigfoot

an imprint of sasquatch books
seattle, wa

Off we go for a walk through the woods! I wonder where the trail will lead and what we'll see along the way.

1

One moose
making a movie.

2

Two raccoons reading
and rocking out.

Ting!

3

Three slugs
sipping cider.

8

Eight coyotes
kayaking and canoeing.

11

Eleven orcas
in an orchestra.

My playing is
particularly
note-worthy.

What a whale-y
nice bassoon!

12

Twelve seals surfing!

Let's join our forest friends!

For my wild and
wonderful kids,
Gwen and Theo! —BC

To Laura, Erin, Darin,
Sunny, and Story. Thank
you for letting me join
your home to complete
this book! —JB

Copyright © 2020 by Ben Clanton
Illustrations copyright © 2020 by Jessixa Bagley

All rights reserved. No portion of this book may be reproduced or
utilized in any form, or by any electronic, mechanical, or other means,
without the prior written permission of the publisher.

Manufactured in China by C&C Offset Printing Co. Ltd.
Shenzhen, Guangdong Province, in March 2020

LITTLE BIGFOOT with colophon is a registered trademark
of Penguin Random House LLC

24 23 22 21 20   9 8 7 6 5 4 3 2 1

Editor: Christy Cox | Production editor: Jill Saginario
Designer: Anna Goldstein

Library of Congress Cataloging-in-Publication Data
Names: Clanton, Ben, 1988- author. | Bagley, Jessixa, illustrator.
Title: Curious encounters : 1 to 13 forest friends / Ben Clanton ;
   illustrated by Jessixa Bagley.
Description: Seattle, WA : Little Bigfoot, [2020] | Audience: Ages 3–7 |
   Audience: Grades K–1
Identifiers: LCCN 2019052976 | ISBN 9781632172747 (hardcover)
Subjects: LCSH: Animals–Juvenile literature. | Counting–Juvenile
   literature.
Classification: LCC QL49 .C587 2020 | DDC 590–dc23
LC record available at https://lccn.loc.gov/2019052976

ISBN: 978-1-63217-274-7

Sasquatch Books | 1904 Third Avenue, Suite 710 | Seattle, WA 98101

SasquatchBooks.com

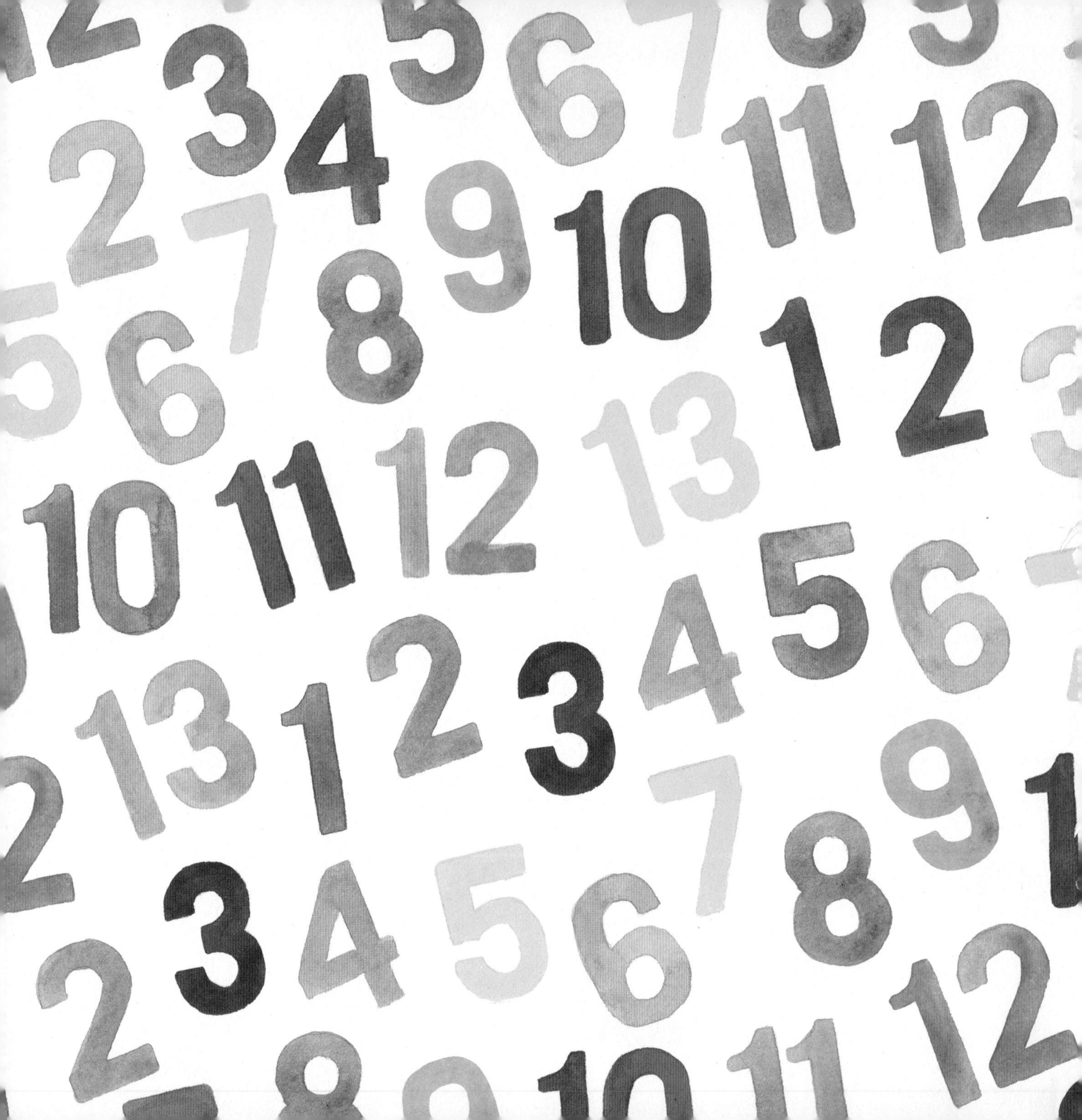